BLEEDING EDGE TECH

THE FUTURE OF VETERINARY CARE

DR. GORDON ROBERTS BVSC MRCVS

TABLE OF CONTENTS

Introduction
Page 4

Chapter 1
3D Printing
Page 6

Chapter 2
Robotics
Page 24

Chapter 3
Virtual and Augmented Reality
Page 37

Chapter 4
Artificial Intelligence
Page 50

Conclusion
Page 60

Reference
Page 62

FOREWORD

The face of veterinary medicine is about to change forever.

There are new developments on the horizon, such as nanotechnology, virtual reality, and artificial intelligence, which will not only transform how we practice veterinary medicine, they will also revolutionise how we study it as a field, and how we train our future veterinarians.

The field of veterinary medicine is about to become more diverse and more specialised than ever before, thanks to this new breed of "*bleeding edge*" technologies.

With this book, I'm going to take you on a journey through these changes and more.

You'll learn about new techniques that, until very recently, belonged only in the pages of a sci-fi novel. We'll shed new light on conditions that have, until now, been a death sentence for our pets. And we'll learn how the pets of the future will live longer, healthier lives thanks to this revolution in pet care.

Are you sitting comfortably? **Good, then we'll begin.**

Dr. Gordon Roberts
The Futurist Vet
January 2017

INTRODUCTION

The field of veterinary care is rapidly changing. As we'll discuss in this book, there are a number of innovations and new developments on the way which will change the face of veterinary medicine forever.

Too new to be called "*cutting edge*", we have chosen to refer to these techniques as "*bleeding edge*" technologies - ideas that we believe will be being introduced over the next few decades.

In chapter 1, we'll talk about what new 3D printing techniques mean for veterinarians, including how they will impact prosthetics and the practice of tissue regeneration.

We'll talk about how veterinary medicine will benefit from these techniques and how they will change what it means to be a vet.

In chapter 2, we will go through some amazing changes to the surgical field with the arrival of a new generation of surgical robots. We'll see how these are already being used in the medical field and how they are just beginning to be used by veterinarians.

In chapter 3, we'll talk about virtual reality and augmented reality, two fascinating areas that will change how we train our future veterinarians and how we prepare for surgeries.

Finally, in chapter 4, we'll talk about one of the biggest changes to arrive on the veterinary scene so far this decade: the arrival of supercomputers and artificial intelligence.

So, what do all of these developments mean for the future of veterinary medicine? In the following chapters, we'll explore those questions and more, giving you a true understanding of what tomorrow's veterinary profession will be like.

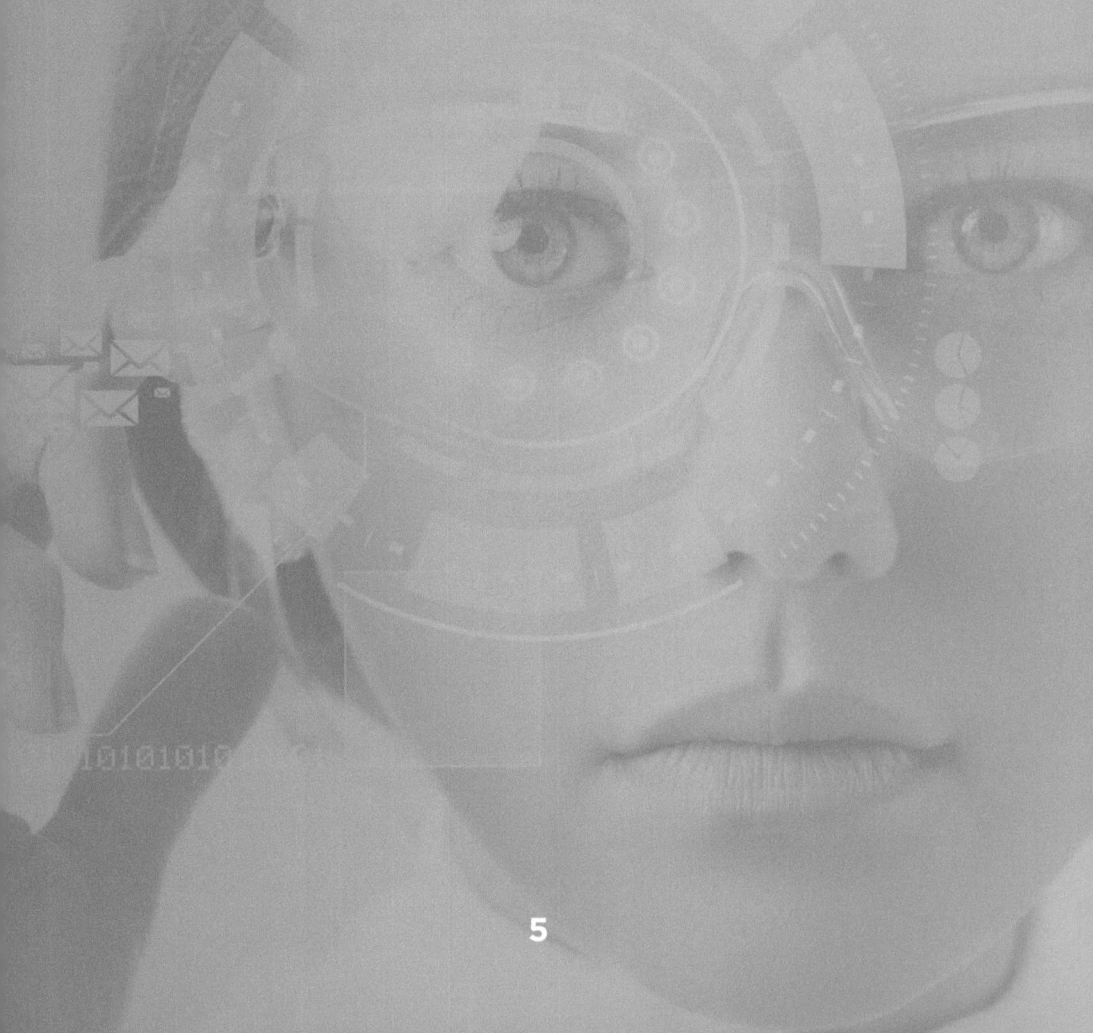

It's important to note that some of these developments are so new that they haven't yet reached the veterinary world. In these cases, we have drawn examples from human medicine to show you exactly what we believe will be possible some day soon, when these technologies trickle down into the veterinary world.

The resulting book will give you some easy-to-understand, real-world examples of what we can expect the veterinary world to look like in the next few decades. I hope you find it useful.

If you'd like to know more about the future of veterinary care, don't forget to look up some of my other books on this subject, which are available on Amazon.

chapter one:
3D PRINTING

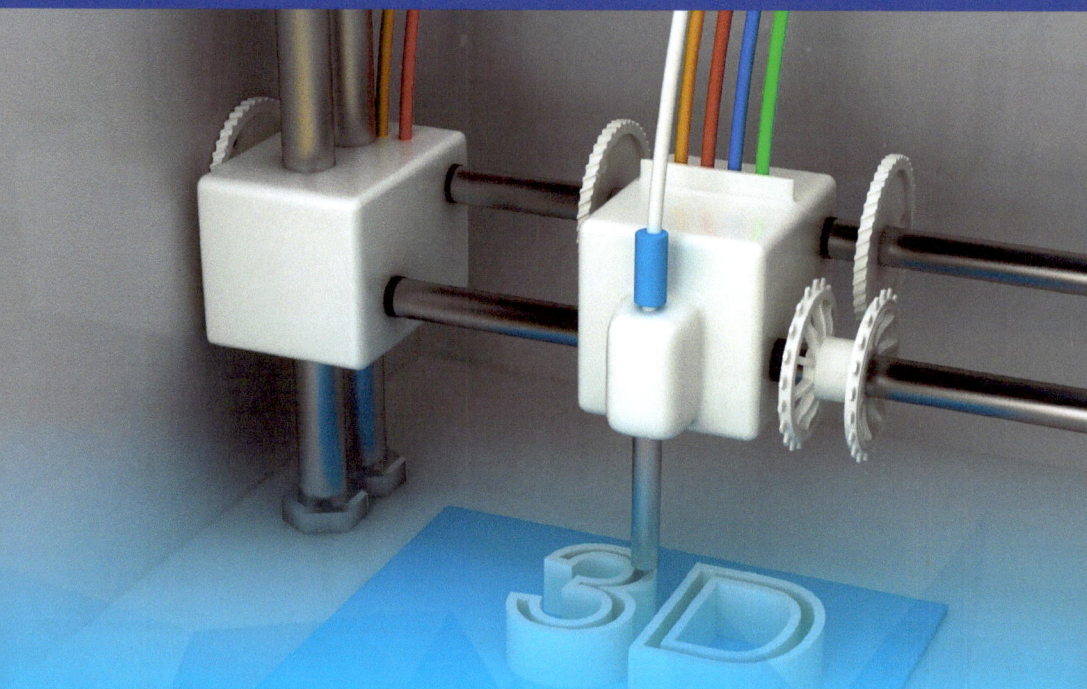

3D printing is currently making waves in the medical field, and we can expect it to disrupt the veterinary industry very soon.

In this first chapter, we'll discuss some of the features and benefits of 3D printing and some ways in which it will revolutionise medicine for both animals and humans.

What is 3D printing, and how does it work?

3D printers are a new generation of machines that can "print" three-dimensional objects in various materials. All these machines need in order to print an object is the computerised three-dimensional design and the material with which to build it.

The machine then produces the design using thousands of tiny layers of the chosen material, which are built on top of each other to produce a solid object. This is called an "additive" process. These layers can be as complex as the design needs them to be, meaning that all sorts of intricate and complicated objects can be produced with them.

To make the virtual 3D design, a CAD (computer aided design) file is required which the 3D printer then reads and reproduces. The CAD file can be made using special 3D modelling software or, if it is an existing object, one can simply use a 3D scanner to make a digital copy of the object.

These 3D scanners are becoming more and more popular in the tech world and it's thought that, in the future, scanning an object onto your computer will be as simple as taking a picture.

Next, the 3D design of the object needs to be "sliced" or divided into thousands of small layers that the 3D printer will then build. A separate piece of software can do this slicing. Once the object has been sliced, it is then ready to be fed into the printer itself, either using a USB stick, an SD card or simply with wifi. The 3D printer will simply read the sliced design and produce the tiny 2D slices to eventually make a 3D object.

The result is a device that can produce anything from ceramic cups to metal machine parts, plastic toys and even elaborate pieces of chocolate. One of the best things about this new technology is that we can use it to produce truly bespoke products; shoes that are printed to fit the dimensions of your own feet perfectly, bike handles that perfectly fit your hands and furniture that fits your home just right. If you accidentally break a treasured possession that is no longer being made in factories, you can simply print yourself a new one. These abilities bring a whole new world of possibilities to how we do things in the future.

The Independent newspaper journalist Andrew Walker writes: "That world, where you can make almost anything at home, is very different from the one we live in today. It's a world that doesn't need lorries to deliver goods or warehouses to store them in, where nothing is ever out of stock and where there is less waste, packaging and pollution."

Some of the most interesting things that people are currently experimenting with are the production of 3D printed organs and 3D printed prosthetics, along with 3D printed medication. All of this makes 3D printing a hugely exciting prospect for the future of veterinary care, which we'll talk about later. First, here's a brief overview of the 3D printing market and where things currently lie. 3D printing: an overview

The idea of 3D printing has been around for quite some time. However, until recently, the practice has been something that was used for producing small trinkets rather than important objects and tools. In the early days, people were sceptical as to whether 3D printing would ever become something that was used widely in manufacturing.

Fast forward to today, and lower costs and advances in technology have put 3D printing back on the radar for large manufacturers and innovative companies. A more agile design process coupled with the fact that products can now be made quickly has meant that 3D printing is rapidly on the rise.

According to a recent study, the market for 3D printing has grown 30% each year from 2012 to 2014. In 2016, it was estimated to become a $7.3 billion market.

A recent report by the company UPS stated that the US has the largest share of the 3D printing market, at 40%, with Europe closely behind at 28% and the Asia/Pacific region at 27%.

However, the 3D printing market still has a long way to go, and currently represents just 0.04% of the global manufacturing market. This is forecast to rise to 5% in the coming years.

There are enormous opportunities out there for 3D printing, not only for manufacturing prototypes but for making functional parts and visual aids.

Industries currently using 3D printing

To give you an idea of how widespread 3D printing is becoming, and how specialised it can be, here are some areas where this technology is already in use (note: this list is not exhaustive):

Automotive industry

The car manufacturer BMW uses 3D printing to make lighter tools for its assembly line. Ford uses it to create complex moulds which are faster and cheaper.

Aerospace

3D printers are being made for use in the international space station. A company called Lockheed Martin is 3D printing satellite parts, with a 48% reduction in cost.

Consumer and Retail

Nike is using 3D printing for its FlyKnit shoes and has managed to reduce waste by 80% in the process. The cosmetic company L'Oréal uses 3D printing to produce living skin in the lab, which it uses for product testing.

Conservation

The biotech startup Pembient has managed to print artificial rhinoceros horns in an attempt to reduce poaching of real life rhinos.

Medical and Healthcare field

A non-profit called E-Nable gives people access to low cost 3D printed prosthetics. A company called Stratasys is partnering with medical institutes to print replicas of human hearts to assist with surgeries. As you can see, 3D printing really is the future of manufacturing, and it is already being used successfully in several industries.

Source: *BSR issue brief: 3-D Printing Sustainability Opportunities and Challenges, November 2015*

What are some advantages of 3D printing?

Producing prototypes

The production of design prototypes is something that 3D printing excels at. That's because, when printing a prototype, there is no need for extra tools to make particular parts.

The 3D printer cuts out the need for setting up time-consuming assembly lines, as well as the need to equip them with specific tools to produce individual parts. Everything is produced by the 3D printer, eliminating the time it takes to set up traditional manufacturing methods.

If the prototype in question doesn't work out, and a design change is needed, this can easily be programmed into the software that the 3D printer uses and the prototype can simply be printed again with the new additions included. This is much simpler than retooling an assembly line and creating new parts the conventional way.

Preventing waste

3D printing only uses the exact amount of materials required, and as such it is extremely efficient at producing items without waste. This is a much smarter way to produce items than, say, using a mould that has to be filed with material, or cutting shapes from a large sheet of metal. The 3D printer only uses what it needs, leaving no costly waste materials.

This makes it a very attractive option for manufacturers looking to produce goods cheaply. Another key reason that 3D printing is cheaper is because it cuts out the resources and people required by a traditional assembly line. To produce an object, all that is needed is the design and the material, making the entire process far simpler and more cost effective.

What are some of the disadvantages?

Printing speed

The current speed of 3D printing could be considered something of a problem when it comes to large scale production. Whilst 3D printing cuts out the need for time-consuming assembly lines, the actual time it takes for an item to be printed can be hours.

This is due to the fact that 3D printing creates objects using tiny layers of material which are built up on top of each other over time. An example is a pair of nutcrackers which can take up to three hours to produce with 3D printing. A traditional assembly line could produce hundreds of nutcrackers in that time. This shows us that, unfortunately, 3D printing can't currently compete with traditional manufacturing lead times when it comes to mass production. However, this is something that engineers around the world are working on.

Materials available

Not all materials are currently suitable for 3D printing. In fact, a new 3D printer usually has to be made for each new material it is going to print with. The melting point of the material is a big factor in whether it Is viable, and as a result we have only recently figured out how to work with certain metals.

Scale

The nature of 3D printing means that we are currently limited to producing parts of a certain size. Very large parts are something of a problem, due to the size of 3D printers themselves.

Very large parts can be produced in smaller sections and then later assembled, but this defeats the ease and purpose of not having an assembly line to begin with. This piece-by-piece assembly might also create stress points in the design, where a solid build would work more efficiently.

The points above show that 3D printing is not something that we can readily use for mass manufacturing just yet. However, work is underway to change that and faster, more efficient 3D printing methods are currently being investigated.

In fact, it's estimated that a third of all manufacturers are currently using 3D printing in some way. Whilst 3D printing isn't suitable for mass production at the moment, it is perfect for

producing bespoke, one of a kind items such as the ones the veterinary and medical world needs. We are going to talk about these next.

3D printing and the veterinary world

It's time to look at what this amazing new technology means for the veterinary world. In many cases, new treatments being tried on humans will later be developed into veterinary techniques.

This means that in order to predict the future of veterinary care, we must also look at what is happening in human medicine. In the case of 3D printing, this means looking at two major areas: prosthetics and tissue engineering.

Prosthetics

3D printed prosthetics are, quite literally, changing the face of modern medicine. This is due to this technology's amazing ability to build completely bespoke parts that fit the wearer's own unique anatomy. This is excellent news for veterinary medicine; just imagine how different the needs of a horse are to, say, those of a baby rabbit.

Animals come in all shapes and sizes and the arrival of 3D printing means we can finally accommodate our beloved pets. There is a huge need for intelligent prosthetic solutions in a field where amputation is often seen as the kindest, safest and cheapest option for injured animals.

Before we get into more examples from the animal kingdom, let's look at what is currently happening in 3D prosthetics in general. In the US alone, it's thought that over 200,000 amputations are performed every year. This means there's a huge demand for prosthetics in the medical field. The arrival of 3D printing has revolutionised the world of prosthetics in many ways, including:

Cost: Prosthetics can cost anything from $5,000 to $50,000 and this means that access for those on lower incomes can be something of a problem. 3D printing is a cheaper way to produce prosthetics and, once it becomes widely available, it will mean more people living with limb loss can get the mobility they need to lead fuller lives. The same can be said for the animal kingdom too. There are currently very few people who specialise in making limbs for animals and as a result, prosthetics are the exception for our four legged friends, rather than the rule.

Timescales: At the moment, getting a prosthetic limb designed, made and fitted can take anything from weeks to months.

This is because prosthetics need to be made according to the individual's anatomy. Imagine if you could simply print your own, using your very own home printer? This would make replacing old or defective limbs a quick and easy process. We're not too far away from this, as the following examples show.

e-NABLE and access to prosthetics

No discussion of 3D prosthetics would be complete without mentioning the astounding work that is done by the non-profit e-NABLE. In 2011, Ivan Owen created a metal puppet hand to wear to a steampunk (sci-fi) convention. He posted a short video of the hand on Youtube, not knowing that one clip would go on to change the lives of thousands.

Ivan started to receive emails from amputees who were interested in the hand he had created. The first of these was Richard, a carpenter from South Africa who had lost his fingers in a work accident.

Collaborating remotely across the world, Ivan and Richard worked through various designs to come up with a replacement finger. This led to another request from South Africa, this time on behalf of a 5-year-old boy called Liam who was born with no fingers on his right hand.

Inspired, Ivan began to research prosthetic devices and he stumbled on a design that was created in the early 1800s by the Australian dentist Robert Norman. The design was made from whale bone, cables and pulleys and was so well made that it later became the prototype for every e-NABLE 3D printed hand. But Ivan didn't stop there. He realised that little Liam would quickly outgrow his prosthetic hand and would soon need a new one. He started researching the idea of using 3D printing to make the next version. He managed to convince a 3D printer company to donate two 3D printers and taught himself how to use 3D design software.

Ivan decided that instead of patenting the design for the new hand and trying to profit from his creation, he would publish the design as an open source file so that anyone, anywhere in the world could download the design and print their own prosthetic hand. At that point, he hoped that more experienced designers would come along and improve on the design, which was still quite basic and chunky.

Word quickly spread about these new 3D printed prosthetics that anyone could make with a 3D printer. Ivan began to build a community of volunteers around the world who had access to 3D printing. The idea was that anyone who needed a prosthetic could get in touch to find their nearest volunteer, who would then print the required prosthetic. At the same time, designers started joining the project and they began to collaborate and share improved designs as Ivan hoped they would.

Today, the non-profit has created thousands of prosthetic devices and given them, free of charge, to people in 45 countries around the world. This story shows how 3D printing, combined with the principles of open source publishing, can be a powerful force for change. It shows us that with improved access to 3D printing, all kinds of amazing innovations are possible.

Prosthetics and animals

As was the case with little Liam, both people and animals grow quickly, and the ability to simply print a new, larger prosthetic using a 3D printer is something that can truly change amputees' lives. Imagine if we could one day have access to the same sort of service when it comes to producing prosthetics for our pets.

Potentially, the vets of the future could create an open source database with prosthetic designs for all kinds of domesticated animals and breeds. This collaborative design library could then be used by animal organisations and pet owners around the world.

Here are some fascinating examples where 3D printed animal prosthetics are already being used, not only for pets but for wildlife, too:

New prosthetic limbs for Derby

Search for "Derby the dog" in Youtube and you'll see how 3D printed limbs can make a huge difference to an animal's life. Derby, a beautiful Husky-cross, was born with no front paws and deformed front legs, making it hard for him to get around and enjoy life like a normal dog. His luck changed when he was fostered by Tara Anderson, an employee of the 3D design and printing company 3D Systems. At the time, Derby had a cart with wheels to help him move around, but it didn't allow him to run and play with other dogs like he wanted to. Tara and her colleagues decided to come up with something better, using the powers of 3D printing.

They decided to move away from replicating the thin design of Derby's existing legs, as it was thought these could get caught in the mud when he was out on walks. Instead, they came up with a curved design that cradled Derby's leg stumps in rubber holders, whilst the rest of the prosthetic was made with a flexible plastic. To get him used to his new prosthetics, the team made them slightly lower and closer to the ground than his real legs would have been if they were fully formed.

Later on, the design was amended so that he had full height legs, and a new technique was used called selective laser sintering. This uses a laser to sinter a special powdered material and build it into a solid structure.

The design has a subtle bending mechanism, just like a real knee. The prosthetics themselves took a few hours to print, using an industrial-grade printer called the ProJet 5500X. With his new legs, Derby can now run and play like the other dogs, and he regularly goes for two hour runs with his new family.

A 3D printed beak for Beauty

In 2005 in Alaska, Beauty the magnificent bald eagle lost the upper half of her beak when she was shot in the face by poachers. Without this vital part of her beak (the upper "mandible") she could not feed properly, protect herself or preen her feathers. She had little hope for survival. Thankfully, she was rescued and brought to the Birds of Prey Northwest shelter.

A mechanical engineer who knew about 3D printing stepped up to help. He made a mould of Beauty's beak and 3D scanned it. He tweaked the design and then printed it using a nylon-polymer material. The finished prosthetic was then attached to Beauty's existing beak using a special titanium mount, allowing her to eat and drink again.

Over the years new 3D models have been made, allowing new iterations of her beak design to be printed according to her changing needs. Recently, there has been more good news: the remainder of her real beak has shown signs of growth and this has pushed the prosthetic beak off. Beauty can now consume certain foods by herself.

Prosthetics and the future

The ability to easily and cheaply create prosthetics for our animals is wonderful news for pets born with disabilities and pets who lose limbs due to injury. Currently, these disabled pets are often abandoned in shelters and can be difficult to rehome due to their special needs.

Getting prosthetics is difficult and expensive, and usually owners resort to makeshift wheel carts or homemade devices which are often impractical and ill fitting. With the new advances in 3D printing, these situations could become a thing of the past and, in the same way as humans, an animal with a disability will be able to enjoy a new level of mobility and a greater quality of life.

As we have seen, the arrival of 3D printed animal prosthetics is already disrupting the veterinary field and once it becomes commonplace, it will mean a whole new suite of services and skills will be needed on a commercial level. Vets will soon begin to acquire their own 3D printers or form partnerships with 3D printing providers. They will collaborate closely with designers and manufacturers in order to produce prosthetics that perfectly meet an animal's needs.

This new specialism is likely to be offered as an option during veterinary training. We can also expect to see a proliferation of specialist prosthetics providers who will focus solely on serving the veterinary market. In short, it's an exciting time for vets who are interested in this area of expertise, and for animal lovers who want to improve their pets' lives.

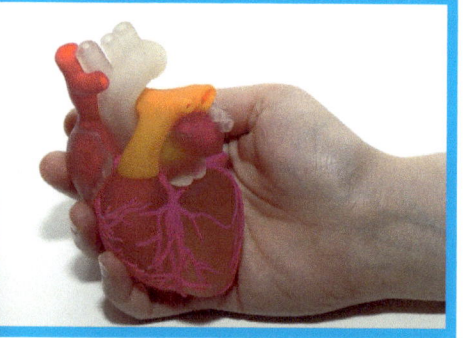

3D printed organs

Bioprinting is another area in which 3D printing is paving the way for the future of veterinary care. 3D printing is currently being used to produce human bones such as hip replacements, allowing for completely bespoke designs.

Scientists are also currently in the early stages of being able to produce entire replacement organs using stem cells and 3D print technology. Since organ transplants happen rarely in the veterinary field, the arrival of these new techniques could change everything.

We may, someday soon, be able to order replacement organs such as hearts, livers and kidneys, for our beloved pets - no matter what shape, size or breed they are. Since animals are already being used for these early experiments, we know that there is an excellent chance that this will become a viable option for our four legged friends.

This isn't the only good news for animals; once we have developed the ability to produce human organs, the need to test drugs and products on laboratory animals, which is now widely recognised as inhumane, will be greatly diminished.

How does organ printing work?

The idea of printing organs works much the same way as printing prosthetics; in order to create an exact replica of the organ needed, the patient's existing organ would need to be scanned using a CT scanner or MRI, and this scan would need to be converted to a CAD file that could be uploaded to the printer. If this organ is damaged or unhealthy, the design will need tweaking to get it to a healthy shape and size.

A 3D printer is used which has nozzles that squirt out the chosen material (in this case, a form of biological "ink"). This bio material is either made up of cells from the original organ, or stem cells, which are harvested from the body and then grown in a petri dish. They are then combined with a bonding agent or gel which keeps everything together. This agent is carefully selected so that the body doesn't reject it. During the printing process, the material is then layered onto a platform until the desired structure is achieved.

Once the 3D printer has done its work, the finished organ then needs to be incubated in a vessel which mimics the conditions of the human body. During incubation, the cells contained in the bio material will fuse together and begin to work as a real organ would.

Challenges

There are a number of challenges that currently lie in the way of us creating and using these artificially grown organs. These include choosing the right material to print with and getting it to live and grow properly outside of the body. Another major challenge is getting the printed organ to behave like a real, functioning organ once it is transplanted.

According to scientists, this is an altogether bigger achievement than simply printing the cells and growing them in the required shape and size. With this in mind, it seems that the real work begins after the organ has been printed. Lastly, the software we reply on currently is not advanced enough to produce truly detailed models of the more complex organs in the human body.

Scientists are working on these obstacles as we speak. Athough it will be another decade or two before we have functioning organs, we are already at the stage where patches of tissue from these organs can be used for drug testing and other experiments, which is a major innovation in itself. It is also hoped that these organ tissue patches can eventually be added to living organs and will help to repair or regenerate them, for example in a diseased liver.

Progress in organ printing

The process of printing using cellular material first began in 2003, when a man called Thomas Boland patented a new process using inkjet printers.

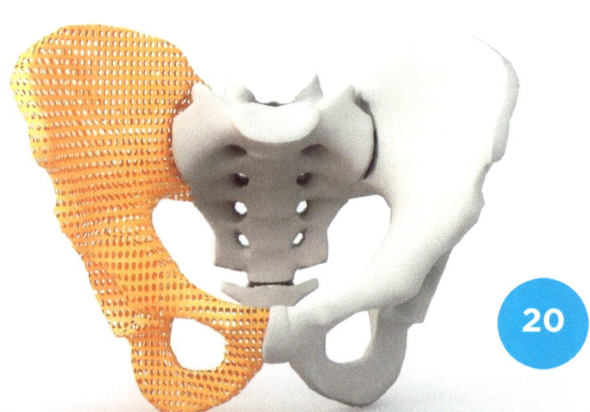

This was a system where the printer was used to deposit cells onto organised 3D matrices.

This new process evolved into what we now call "bioprinting" and, instead of using 3D matrices to print onto, scientists have more recently began to focus more on tissue and organ structures. As well as drop-based inkjet printing, a new method of 3D printing has been developed, known as extrusion bioprinting.

This method uses a mobile printhead to deposit a particular material. This technique allows for greater cell densities and is a gentler, more controlled process than inkjet printing.

Although cellular printing material was only developed in 2003, the idea of 3D printing has been around since the early 1990s. Not long after that, medical researchers began to ponder the question: if all sorts of other objects could be produced with 3D printing, why not organs? This was something that could be made possible if the existing 3D printers could be made to print bio materials instead of plastics.

By the late 1990s, scientists had developed some techniques to do just that. In 1999, scientists at the Wake Forest Institute for Regenerative Medicine managed to 3D print a scaffold of a human bladder. Once the scaffold was ready, they coated the structure with patient's cells, and from there they were able to succcessfully grow it into a working bladder.

An organ like a bladder is easy to produce, firstly because of its relatively simple, hollow shape and secondly because the bladder only contains two types of cells. Organs such as kidneys and livers are far more complex in terms of their cell types and structures. This amazing advance in medicine led to the practice of bioprinting. Not long afterwards, in 2002, scientists produced a miniature kidney that could filter blood and produce urine in an animal model.

Since these developments, scientists have produced miniature models of all sorts of organs, from kidneys and bladders to skin, cartilage, ears and urethras.

Many of these organs are functioning, but at this point they have only survived in the laboratory for a maximum of 40 days. We are still quite a long way away from producing an organ that can survive a lifetime.

So, how long will we have to wait before 3D printed organs become available for transplant patients? A while, it turns out. Dr Anthony Atala, director of the Wake Forest Institute for Regenerative Medicine, is an expert in these regenerative techniques, having delivered TED talks on the topic in 2009 and 2011.

Not only did Atala's team create the first lab-grown organ to be successfully transplanted into a human (a bladder), his team are now working on experimental fabrication technology that can "print" human tissue on demand.

In a 2015 interview with the Huffington Post, Atala told journalist Ryan Grenoble:
"Science is unpredictable, so it is impossible to make predictions. But I think we can safely say that the timeframe required to routinely print and implant complex organs is decades, rather than years."

Nonetheless, the amazing techniques involved in producing 3D replicas of organs are already being put to good use. 3D printing in general is making complex surgeries easier by allowing surgeons to have printed replicas of the organs they will operate on, in order to practice and select the right tools in advance.

Surgeons have also used 3D printed implants such as "sling" devices to do things such as hold collapsed airways open. So, even though we'll have to wait a long time until bio printed organs make it to hospitals worldwide, 3D printed organs made from synthetic materials are already becoming commonplace.

What this means for the veterinary field

- We can expect these developments to take slightly longer before they disrupt the field of animal medicine. However, it is highly likely that they will do so once the technology and the medicine to do so has been sufficiently developed. This means that we'll hopefully one day live in a world where:

- Our pets live longer, healthier lives due to the availability of printed organ transplants. This will mean we can effectively cure chronic diseases such as liver failure, diabetes, heart disease and more. These conditions, which today mean a death sentence for our pets, will some day be remedied through transplant surgery, or even through regenerative techniques such as tissue engineering.

- Animals that are the victim of burns, skin diseases or serious wounds will be able to benefit from lab printed skin which will then be applied to our pets in order to encourage new skin to grow and heal.

- Pets that suffer from issues such as luxating patella, a major issue with certain dog breeds, may some day soon be treated with 3D printed knee caps to replace their defective counterparts.

- Arthritis, often a cause of severe pain and discomfort in our beloved pets, could also become a thing of the past with the arrival of lab grown cartilage for veterinary purposes.

- These are just a few examples of how 3D printing organs, skin and bones will one day bring huge benefits and advances to veterinary medicine.

chapter two:
ROBOTICS

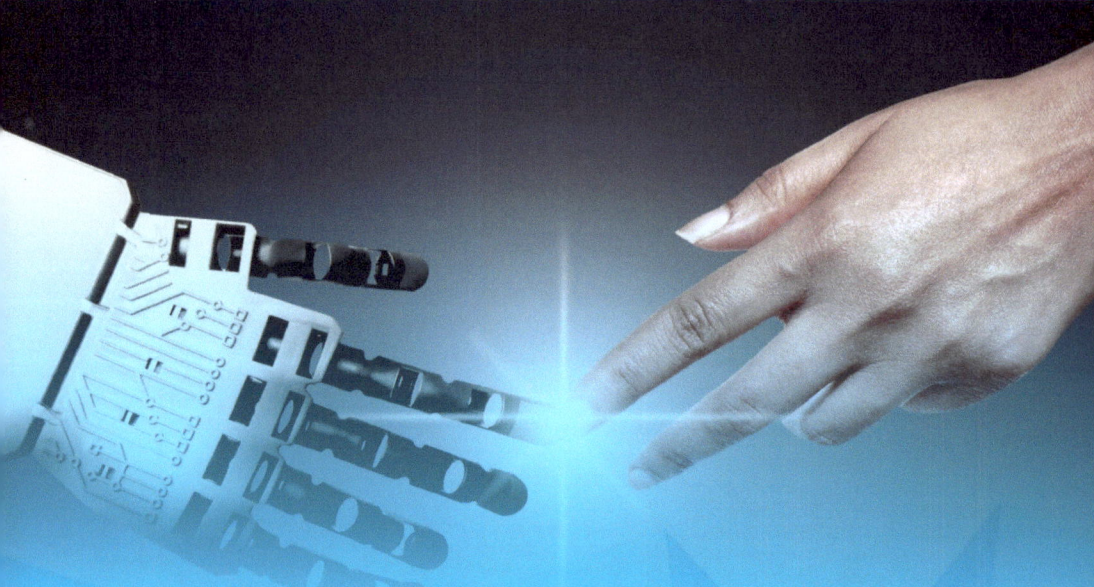

Robotics are currently being used in several areas of medicine, for example in the area of assistive technologies, where robotic mobility aids are improving the quality of life of disabled and elderly people. In the area of rehabilitation, robotics are also being used to help with clinical therapy and neuro-motor rehabilitation. They are even being used to help train doctors and surgeons. The area of medicine we're going to concentrate on first, however, is something that's currently disrupting the medical field, and will some day soon revolutionise veterinary medicine: robotics for surgery.

How will robotics change veterinary medicine?

Put simply, they will mean vets can operate on all manner of creatures with a level of precision that has been unheard of until now. Robot hands are far more stable than our shaky human hands can be. We'll be able to operate on much smaller animals who would perhaps otherwise go untreated. The margins for error will be greatly reduced with these new techniques, and surgeries will be less invasive and less time-consuming.

Lastly, robotics will allow us to work on animals remotely, through computer controlled devices, meaning greater access to specialist veterinary care. All of these are hugely exciting prospects for anyone in the veterinary field.

Why use robots for surgery?

Robots are currently taking over human medicine, and with good reason. In many ways, we can rely on them to perform far better in surgeries than their human counterparts. For one thing, they don't suffer from fatigue or lapses in attention spans like human surgeons do, making them ideal for performing long, painstaking operations.

They aren't swayed by emotions, either, for example they don't get stressed if an operation proves more complex than initially intended. A robot isn't emotionally involved in the procedure, so it doesn't get worried about catching infections and it will never get squeamish at the sight of too much blood.

Crucially, robots are also resistant to radiation. They are easy to sterilise and use super-human precision thanks to special sensors. Being less invasive, robotic surgeries don't cause huge open wounds and, for this reason, they tend to be cleaner and the rate of infection in patients is much lower.

All of this brings a great many benefits to patients undergoing robotic surgeries, including:
• Shorter recovery times and shorter stays in hospital

- A lower risk of death due to infections and other complications
- Less scarring and faster healing
- Less pain after surgery

Using surgical robots: the current challenges

All of the above may sound fantastic, but we are still a long way from trusting robots to carry out surgeries independently. Human surgeons are still preferable for many reasons. For one thing, a robot can't apply critical thinking or judgement to a particular medical case. They simply can't think for themselves.

Surgeries often turn out to be more complex than they first appear (for example, the surgeon finds that they need to work close to a major artery in order to remove a tumour), and unfortunately our robotic friends are simply not cut out to make difficult decisions.

Here are some other issues with using surgical robots:

- Currently, a different robotic machine is needed for each type of surgery, and robots cannot be programmed to perform several different surgeries in one day. For example, a completely different machine is needed to perform heart surgery than, say, a hip replacement.

- Mechanical or technical issues such as power cuts, computer crashes and other electronic failures present a huge risk that there is no way to mitigate currently

- Robotic machines currently cost millions of dollars to build, ship, install and maintain. Hospitals simply don't have the funds to use them widely. In addition to this, surgeons and nurses also need to be trained how to use this equipment, taking up even more time and money.

- As mentioned earlier, each robot can currently only perform one type of surgery, meaning each hospital would require several machines if it were to switch to an all-robotic model of surgery

- Robotic surgical machines currently take up a huge amount of space when compared to a single human surgeon, and space is a resource that many hospitals do not have at their disposal

How robotics has developed

You might be surprised to know that we've been using robots in hospitals for years now. Here is how the field of robotics has developed over the years:

In 1983, doctors created the first surgical robot, known as the "Arthrobot". This robot was developed for orthopaedic surgery, and was designed to manipulate and position limbs during surgery, in response to a voice command by the surgeon.

Dr James McEwen, who was part of the team that developed the robot, told The Medical Post in 1985:
"The surgeon no longer has to do two jobs at once - that is, manipulate the joint and perform the procedure.Alternatively, the surgeon no longer needs a human assistant to position and hold the limb while he operates. Holding a limb in place for long periods of time can be very fatiguing. The robot doesn't get tired and it doesn't get bored. The whole idea behind the robot is to reduce the labour intensiveness of certain surgical procedures."

Surgical robots continued to be developed through the 1980s and 1990s, including the PUMA 560 in 1985, which was used to place a needle for a brain biopsy.

In the late 1990s, robotic technology began to be used in heart surgeries for the first time. For example, in May 1998, the first robotically assisted heart bypass was carried out by doctors in Germany. The following year, the same operation was carried out in the USA at the Ohio State University.

Further breakthroughs happened in 1999, when the world's first beating heart coronary artery bypass graft (CABG) was performed robotically.

These first-time robotic operations continued throughout the 2000s, across a wide range of different surgeries including bladder reconstructions, microsurgeries, and kidney transplants. By the end of the decade, doctors were not just performing robot-assisted surgeries, they were also overseeing all-robotic surgeries without human intervention.

To give you an example of the types of robotic devices being used in the medical field, here is a table showing some of the machines in use currently. This is taken from the publication BMJ, formerly the British Medical Journal:

Name of robot	Type of robot	What it's used for
Endoassist	Active camera	MAS camera manipulation (synchronised to surgeon's head movements)
Fips endoarm	Active camera	Minimal access surgery camera manipulation (finger ring joystick controlled)
AESOP	Active camera	Minimal access surgery camera manipulation (voice controlled)
Minerva	Active surgical	Stereotactic neurosurgery
Acrobot	Semi-active surgical	Prosthetic knee implantation
CASPAR	Active surgical	Prosthetic knee implantation
Robodoc	Active surgical	Prosthetic hip implantation

Name of robot	Type of robot	What it's used for
Probot	Active surgical	Resection of benign prostatic hyperplasia
Inch-worm	Active autonomous	Colonoscopy
Da Vinci	Master-slave telemanip-ulator	General, cardiothoracic, and gynaecological surgery

Here are some real-world examples of how the medical field is currently using robotics. Many of these techniques have already been around for years. However, we will need to wait for the technology to become much cheaper and more widely available before we can use it in the veterinary world. When this happens, our pets will reap the benefits.

The da Vinci robot

No discussion of surgical robotics would be complete without mentioning da Vinci, a robotic system that was created by an American company called Intuitive Surgical. This machine enables surgeries to be performed through making small incisions, and is a fantastic example of what we can achieve with robots in the medical field. The da Vinci system translates the movements of the surgeon into small, precise movements made by tiny surgical instruments inside the patient's body.

These instruments include scalpels, scissors, and bovies. The instruments are controlled remotely with the help of a laparoscope - a thin tube with a tiny camera and light at the end. The images that the camera picks up are transmitted to a video monitor which the doctor uses to guide his movements, using a special master control console.

The genius of the instruments lies in the fact that they have a jointed wrist design, which exceeds the natural range of motion of the human hand. The surgeon is always in control of the da Vinci system - that is, it never operates independently.

The surgeon sits at a special console with everything but the screen itself shielded from view, so as to have total concentration. This is a huge improvement on traditional laparoscopic surgeries, where the surgeon used to have to stand for long periods, using long instruments which did not bend like wrists.

So far, this technology has been used on 3 million patients worldwide, and has revolutionised the surgical field.

It has been used for a wide range of surgeries, including:
*   Cardiac surgery
*   Colorectal surgery
*   Gynaecologic surgery
*   Head and neck surgery
*   Thoracic surgery
*   Urologic surgery

The da Vinci system was approved by the USA'S Food and Drug Administration in the year 2000, and since then it has been used in hospitals all over the world. In June 2014, there were 3,102 da Vinci systems in operation around the world.

The most common surgeries it performs are hysterectomies and prostate removals. To truly see it at work, it's a good idea to watch Catherine Mohr's excellent TED talk: Surgery's Past, Present and Robotic Future (see www.ted.com).

However, the da Vinci system is not without its critics and challenges. It has been described by critics as somewhat difficult to learn, with no real proof that it is more effective than the traditional laparoscopic surgery.

Cost is another factor - it's $2million price tag makes it difficult for cash strapped hospitals to procure. Another disadvantage is that this robotic technology is said by some to dissolve the "creative freedoms" of the surgeon. However, the machine's worldwide use, and its success rate, speaks for itself and its safe to say that the da Vinci will be around for a long time to come. Surgical robots in veterinary care

The exciting news about all this is that robots are just beginning to be used for veterinary surgery. In April 2015, eight year old lion Leonardo had surgery to remove an adrenal tumour.

Using a device called the Telelap ALF-X, vets at the Veterinary Hospital of Lodi, Italy, were able to perform the life-saving surgery. Leonardo had been showing symptoms of an endocrine disorder, so he was moved to Milan for a CT scan.

Vets discovered that he had a large tumour in the left adrenal gland, which they decided needed to be removed. This was the first time that any lion had presented with adrenal cancer, so the vets wanted to proceed with caution.

They realised that traditional surgery by human hand would not only be a long and stressful operation, it would also bring the risk of major tissue damage and would mean a long and painful recovery for Leonardo. There was a high chance of complications such as infection, from the large wound that would result from the surgery.

To mitigate these risks, the team decided they would perform a minimally invasive robotic surgery. Much like the da Vinci, the Telelap ALF-X robot works with robotic arms that the surgeon controls remotely, from a console. Surgeons made three small 3cm incisions and were able to see a 3D high definition view of Leonardo's internal organs. This state of the art software meant that surgeons could get a proper "feel" for the tissue they were working with. Eye tracking software also helped to move the camera and take shots where needed. This was the first robotic operation on an animal to date, and it is hoped that many more will follow.

Examples of robotics in other areas of medicine

We have talked a lot about surgical robotics and what this might mean for the veterinary world. But surgery isn't the only area of medicine to benefit from the genius of robots. Let's look at some other ways that robots are helping humans in the medical field.

Diagnosing diseases

When it comes to diagnosing illness, things are rarely straightforward. With human error, there is a huge risk of missing vital symptoms or information that could otherwise save lives. For example, sometimes trace elements in the blood stream are so small that they go undetected. Unusual growths or masses might be too small to show up properly in an x-ray. As well as this, some diagnostic procedures are just too invasive to be worthwhile when it comes to early detection.

The solution to all this? Robotics, of course. Robots can be far more precise in a way that humans cannot always be. Robots can now do amazingly precise and non-invasive things, such as taking biopsies, or using ultrasound to provide 3D images about unusual masses found in the body. Some robots can even detect tumours or aneurysms in the brain.

Rehabilitation

Patients with serious injuries often need months or years of physical therapy and rehabilitation in order to regain their lost mobility. Robots can help with this by providing special robotic devices to aid people in regaining strength and muscle function.

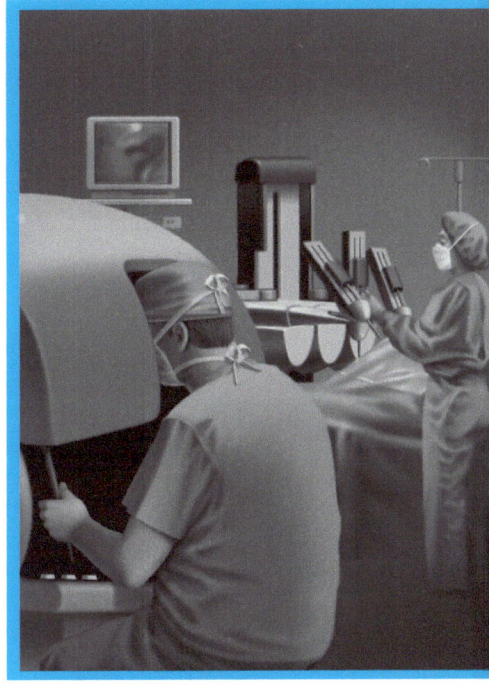

Training doctors

Medical students need to practice procedures on real-life people before they can become certified doctors. However, often there is far too great a risk to the patient. Robot patients have been created to fill this need instead. The robots have a beating heart, dilating pupils and can even breathe. This allows students to practice things such as fitting an IV, taking blood pressure and even talking with the robot about its symptoms.

Moving hospital supplies

The Aethon TUG can transport items such as medication, bed sheets and food from one area of a hospital to another. According to its manufacturers, this system is far more efficient than its human counterparts as it doesn't get distracted on deliveries. The system can be attached to various hospital trolleys and is very adaptable to different needs. This system reportedly saves money, too. One of these robots working two shifts every day saves the work of almost three full-time employees, costing less than one full-time employee. As such, thousands of these "service robots" are set to arrive in hospitals over the next few years.

Nanorobots

We've discussed the field of robotics as it is currently, and seen how it is beginning to be used in the veterinary field. But what's the next step for robotics? **The answer is nanorobots**.

In 1989, IBM scientists proved that individual atoms could be manipulated by using an electron microscope to position 35 xenon atoms to spell out the letters of the company's name. It is hoped that the same technology will, in the future, be used to produce tiny, diamond-like machine parts which will be used to assemble medical nanorobots.

A collective of scientists called The Nanofactory Collaboration, founded in 2000, is working to do just that. They hope to create the first working nanofactory that will create medical nanorobots within the next two decades.

What are medical nanorobots?

At the moment, these nanorobots are just a concept. However, it is hoped they will become a reality in the near future, because they will be capable of some amazing feats due to their tiny size. Medical nanorobots will be able to squeeze through tiny capillaries in the human body, as they will be smaller than a red blood cell.

It is thought that they will be the size of bacteria, with a construction that's made up of thousands of molecule-sized mechanical parts, which might resemble things like gears, ratchets and bearings. These miniscule designs will need motors to allow for movement, as well as mechanical arms and legs which will act as manipulators. Nanorobots will also need some form of power supply, sensors and a computerised system so that we can connect to them and control them remotely.

What will nanorobots do?

Nanorobots will be able to perform some hugely important procedures, and will change the face of medicine. They would, for example, act as a faster, more effective alternative to antibiotics. A "microbivore" nanorobot has been envisioned, which would act in the same way as our white blood cells do, seeking out and digesting things like bacteria, viruses and fungi in the blood.

Germs like these would stick to the nanobot and eventually the nanobot would ingest them, mincing them into harmless tiny molecules in a matter in minutes. The genius of this lies in the fact that no bacteria or virus would have time to develop a resistance to these nanobots.

A patient would need to be injected with about 100 billion of these nanorobots in order for this treatment to work. However, the treatment itself would take just a few hours - a far shorter time period than it takes for antibiotics to take effect.

Once finished, eliminating the nanorobots from the body is simple. The doctor would simply broadcast an ultrasound signal to direct the robots to the kidneys, where they would be flushed out harmlessly in the urine.

This type of procedure would also mean amazing things for cancer treatment. It is thought that these nanorobots could be programmed to seek out and destroy the smallest clusters of cancer cells, well before they grown and spread throughout the body.

A second type of nanorobot, known as a "chromallocyte" could be programmed to seek out diseased cells and extract the defective chromosomes from them, replacing it with healthy ones. These new chromosomes would have been manufactured in advance, using a desktop nanofactory.

Nanorobots would simply be injected into the body, where they would travel to the target cell and enter the nucleus. Once there, it would replace the required chromosomes and leave the body.

This means we would be able to completely remove the risk of a patient inheriting a chronic illness for which they were genetically predisposed. Taking this idea even further, it would mean we could even reverse the aging that takes place in individual cells, effectively reversing the aging process.

Nanorobots and the veterinary field

Nanorobotic medicine would be a huge revolution for the veterinary industry. Since our beloved pets can't tell us that they're in pain, it would finally mean a way for vets to catch chronic disease early on, before it develops into life-threatening illness.

This, in turn, would mean that our pets and, indeed, wildlife in human car could live a much longer, less disease-prone life. In the pet world, it would mean we could easily "weed out" hereditary diseases that plague certain breeds of dog, for example.

Just as we would be able to reverse the aging process in ourselves, eliminate hereditary diseases and prevent cancers spreading, we could also do the same for our pets. Amazingly, we could one day find ourselves with a pet dog who lives as long as we do, thanks to these advances in medical science.

chapter three:
VIRTUAL& AUGMENTED REALITY

Virtual reality and augmented reality might not be the first terms that spring to mind when we consider the veterinarians of the future.

However, as you'll find out in the following chapter, these two technologies are set to make some huge changes in veterinary training, some of which are already being trialled.

Virtual reality and its possibilities

Technologists have been attempting to create VR (virtual reality) experiences that simulate human senses for over 60 years now. The earliest of these machines were often bulky devices offering limited content. It is only in the past two decades, however, that major advances in computing and video games have taken place. These changes have allowed us to create VR devices that are less bulky, more compact and more immersive.

Today's devices can actually trick the human brain into thinking that the imaginary world they are presented with is a reflection of reality, thanks to biological circuits and sensors.

Like smartphones, these virtual reality and augmented reality devices are expected to appeal to both consumers and companies alike. This new technology brings with it an endless list of possibilities. For example, the desktop workstations that office workers use every day could become virtual work areas multiple displays and real-time access to company data, regardless of where an employee is located.

An employee could choose to work from any virtual setting they wish - from a jungle to a deserted island. Consumers could also use virtual reality to explore the features and benefits of a company's products within their own personalised environments. These are just some of the ways virtual reality could transform every day life.

Virtual reality vs augmented reality

It's important to note the difference between these two terms, before we go any further. Virtual reality (VR) creates an artificial setting using realistic images and sounds to simulate the user's presence in a virtual world. This world could either be based on somewhere in the real world (i.e. a simulation of Hyde Park in London) or it could be somewhere entirely fictional.

To immerse themselves in this world, the user usually wears a headset called an HMD (head mounted display). This allows them to look around their simulated world and interact with it.

By contrast, augmented reality provides users with a real-world view that is enhanced or overlaid with computer generated graphics or data. For example, menus could appear with important information relevant to that particular moment or setting. Users still need to wear a headset for this - Google Glass is an example of this concept already in use.

The following chart will give you a good idea of these two technologies and where they currently lie in the technological landscape.

	Virtual Reality	Augmented Reality
What it does	Changes reality by placing the user in a 360-degree imaginary world.	Visible world is overlaid with digital content.
Where it stands	Has been around for a long time; most famous example is Oculus Rift. Hundreds of companies are working on prototypes.	Introduced in the formof Google Glass. Now several companies aredeveloping prototypes.
Market opportunities	Videogames, theme parks, entertainment apps,video, collaboration, employee training, simulation exercises	Games, theme parks,simulation exercises, employee training, commerce

	Virtual Reality	Augmented Reality
Biggest players	Oculus RV, Samsung Gear VR, Sony, HTC	Vuzix, Skully, Epson

VR in medicine

Workplaces and shopping habits aren't the only areas of our lives that virtual reality is set to revolutionise. The medical field is also gearing up to use these new technologies. This, in turn, will mean changes for the veterinary world, too.

Perhaps the biggest area where virtual reality will make a difference in the medical world is in training new doctors and surgeons. Virtual reality allows for a safe, secure environment for trainee surgeons to practice certain procedures without the risk of harming patients. In this simulated environment, they are free to try new things and make mistakes, without the stress of it having real-world consequences.

This is perhaps even more useful for veterinary surgeons, who need to know how to operate on a wide range of different animals, often with very different anatomies. These new technologies will also allow students to pause procedures, repeat certain actions, zoom in and out, as well as walk 60 degrees around the patient. This new way to learn offers a richer experience and many more opportunities to absorb new information.

Examples of VR in healthcare

Although the major focus with VR is currently on video gaming, there are a number of organisations who are quietly working on VR solutions for the healthcare market. These examples will help us to see what we can also expect to happen in the veterinary world. Here are some of them:

Immersive Touch

This company are currently working on creating VR enabled surgical training through simulation. As discussed above, this would allow for a truly safe environment for trainee surgeons to learn their craft in. Simulators immerse the student in a digital operating environment, one where they will be able to identify and use tools and devices such as replicas of instruments, surgery specific foot pedals, ultrasound guidance and even endoscopic (keyhole surgery) views.

This will be hugely helpful to veterinary students once it reaches the veterinary field; it will allow vets to practice on a wide range of animals where, in reality, there may be a shortage of specimens to practice on. It will also allow vets to practice in different environments, for example vets who practice on farms will need to know how to work from the field as well as their own surgeries.

Medical Realities

Medical Realities are working on a "virtual surgeon" which will give a 360 degree view of surgery. The company has already captured the very first 360 degree video of a surgery, a laparoscopic right hemicolectomy. These kinds of videos allow for a never before seen level of depth and perspective.

They will be invaluable for surgeons who are just starting out in the medical field. This is another great feature that will hopefully benefit the veterinary world in the near future. It is extremely helpful for a vet to be able to see the patient from all angles during surgery. This will reduce complications and increase success rates.

Deepstream VR

This organisation is concerned with pain relief and how VR can help to alleviate pain. They have developed an app called "Cool!" in order to achieve this. The app surrounds the patient with a beautiful landscape where the seasons change and several different creatures make an appearance. There is no linear pattern - that is, no set start or finish - so the patient can stay immersed in the experience for as long as they like.

Studies carried out on the app have shown that patients can experience a reduction in pain of 60 - 70% during the treatment, and the effects of this can last for up to 48 hours with no adverse affects such as the ones pain relief medication can cause. Will we be trying this one on our pets? Who knows? We currently have very few ways to tell if our four-legged friends are in pain, but with certain chronic conditions or procedures, it is almost a certainty. Any method which will lessen an animal's distress is worth trying, surely.

Virtually Better

Virtually Better is working on something called exposure therapy, which works to alleviate stress and phobias by gradually exposing the patient to the stimuli they find frightening. The idea is that, once the patient has had enough encounters with the stimuli in a safe and controlled environment, they will no longer see it as a threat.

This company's new VR based therapy "Bravemind" has some huge benefits for those with PTSD (post traumatic stress disorder). This exposure therapy may also be useful for vets who need to operate on or handle dangerous animals for the first time. Exposing themselves to these animals prior to handling them in real life will lessen stress and improve performance. Of course, members of the public could also benefit greatly from exposure therapy if there are certain animals they have phobias of, such as snakes, spiders and dogs.

Firsthand Technology

This company is using a mixture of VR and gaming to promote good health practices like fitness, hygiene, managing pain, mental health and behavioural changes. Their "Keep Your Teeth" game aims to encourage kids to brush their teeth whilst also increasing cognition through game playing.

These educational resources could be a vital way to spread the word about health practices in the future. If we were to bring this technology to the veterinary field, vets would have a fantastic way to educate pet owners about certain aspects of pet care. They could also educate owners on specific diseases and conditions, and show them how to manage these conditions at home. On a wider scale, these virtual reality simulations would be a great way to teach children how to properly handle pets and help them learn the very real responsibilities involved in getting a pet. If used on a wider scale, this could potentially decrease the number of unwanted pets abandoned in shelters and improve animal welfare in general.

Augmented reality

We've discussed the various ways in which veterinary medicine could benefit from virtual reality. But what about augmented reality? As we mentioned earlier, this works slightly differently to VR. It involves seeing reality instead of a simulated environment, only the reality you see is overlaid with pictures, words, or data regarding the things in your field of vision. Google Glass is an example of this concept in its infancy. It is thought that this simple yet sophisticated idea will be used in all kinds of ways throughout our daily lives.

Imagine your car breaks down in the middle of nowhere. Perhaps you are late for an appointment or you don't have enough battery on your phone to call the garage. In an AR-enabled future, you would simply take out your smart glasses, launch a car repair app and follow step by step instructions that pop up in your field of vision. With the help of your device, you quickly and easily figure out what's wrong with your car and are ready to take action.

Or, perhaps you're out shopping. You want to know more about a particular dress you've seen. By simply looking at it through your smart glasses, you could find out all sorts of related information such as what other customers paired the dress with, what sizes are available and what fabric it is made from. You could even read reviews from other customers who bought it.

These are just two simple examples of the kinds of things AR could make possible in the near future. In reality, the possibilities for this technology are endless. The definition of AR is "the expansion of physical reality by adding layers of computer generated information to the real environment" (DHL. 2014).

This computer generated information could include text, graphics, video, sound, GPS data, and even smell. This is more than just display technology. This is a natural, real-time interface allowing humans to interact with their environments in a digital way.

How AR works

Augmented reality works in four distinct stages to produce its final output:

Scene capture:

The reality or environment that is going to be augmented is first captured using an AR device. This could be via an HMD (head mounted display) or through a camera of some sort. In the case of Google Glass, it is captured through a set of goggles or glasses with an integrated camera.

Scene identification:

Once captured, the reality environment must be scanned to determine where the virtual content is going to be embedded. There are a number of ways this could be determined, largely depending on what has been captured. For example, visual tags could be used, or tracking technologies such as GPS, sensors, infrared, or laser.

Scene processing:

Once the environment has been clearly captured and identified, the corresponding digital content is requested. This content will come from the internet, or from a database of sorts.

Scene visualization:

Lastly, the user will see their immediate environment overlaid with the virtual information that the AR device has provided.

AR in everyday life

Very few AR solutions are currently available to buy in shops. The first iPhone app that featured AR was released in 2009. Most recently, AR has gained a lot of media attention due to the augmented reality game Pokémon Go. This game is largely

credited as bringing AR to the masses. It involves interacting with reality in order to play, thus since its release, hordes of people have been seen running around public parks and streets " looking for Pokémon" with their phones.

This format has proved hugely popular. In fact, by the end of 2016 it is estimated that there will be around 500 million downloads of the game in total. The AR market is expected to grow from $515 million in 2016, to $5.7 billion in 2021. It is thought that the majority of this income will be a) from a license and subscription fee model and b) largely related to smart glasses and tablets (Forbes, 2016).

AR in healthcare and vet care

So, we know that AR is already making waves in the gaming and entertainment world. But what about using this technology for the greater good? AR is also slowly making its way into the world of healthcare. As such, we can also expect it to reach the veterinary field some day soon. Here are some examples of how AR is being used in the medical field:

Google Glass surgery

Surgery is already being performed with surgeons who are wearing Google Glass, and broadcasting the operation live to audiences. However, later on this ability to merely broadcast could turn into more complex features.

For example, AR could allow a surgeon to simply look at a patient and immediately get vital information such as their age, medical history, or heartrate. Shafi Ahmed, founder of the healthcare company Medical Realities, has used Google Glass to perform live operations for around 14,000 students in 32 countries around the world. The idea behind this initiative is that it creates a global community of surgeons and students, and helps students to learn in a more interactive way.

University of Wisconsin

The good news is that Veterinary colleges have already started to use this technology as a training aid. The University of Wisconsin is using the device to film surgeries, as well as rarely seen diseases such as rabies, in order to train students.

UW School of Veterinary Medicine Instructional Designer Tyler Gregory told the news site Channel3000 in March 2014:
"If a particularly interesting case comes in, the faculty clinician will grab the device, document it and then afterward we can show that content to students who may not have seen a rabies case or something interesting like a blockage or a surgery."

Virbac

The animal care company Virbac has already created an app that informs pet owners about their products. You simply open the Blippar app and point it at the product. The smartphone will then display information, tops and videos related to that product. For example, for the flea treatments, owners will be able to see when and where to apply it, with useful diagrams and videos on application. The app will also work when pointed at leaflets, billboard adverts and even fridge magnets. For vets, this is a useful way to give customers more information, which they can then read in their own time.

ARnatomy

AR could be a fantastic way to help medical students to learn anatomy. A company called ARnatomy is working to make this a reality. They are developing an app that uses OCR (optical character recognition) to match words like "femur" to corresponding visuals and information.

They are also creating a skeletal model of bones affixed with augmented reality (AR) targets, providing key information on specific bones and parts. This is something that could translate particularly well to the veterinary world, and having various models of different species could be an invaluable way for vet students to learn.

University of Liverpool

Veterinary students at the University of Liverpool are already benefitting from the kind of augmented reality techniques mentioned above.

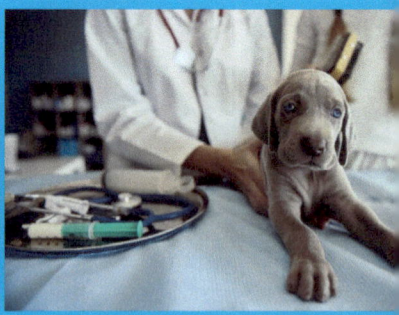

"The school has a very active technology-enhanced learning team that is always looking for new ways to engage learners and help them get the most out of their learning experience here," said Avril Senior, a lecturer at the university's school of veterinary science, told Horse and Hound in January 2015.

The students are using their smartphones to learn about animal anatomy. When held up against anatomical drawings, student's smartphones reveal a 3D image of an equine heart. Phones can also be held up to the operating theatre doors to reveal a video of a horse in the operating theatre. The technology is currently being trialled, and it is hoped that it will, some day soon, be introduced to the school's new curriculum.

AccuVein

Having blood taken can be a very unpleasant experience for some, especially if multiple attempts are needed to find a vein. AccuVein is working to change this. It is a handheld scanning device that projects over the skin and shows nurses where veins are located in patient's bodies.

The device makes finding a vein on the first attempt 3.5 times more likely. It has already been used on over 10 million patients. This is also something that could be hugely useful to veterinary medicine. As species and breeds differ greatly in their anatomy, having a guide such as this would prove especially useful to veterinary nurses and vets.

chapter four:
ARTIFICIAL INTELLIGENCE

So far, we've talked about 3D printing, robotics and virtual reality, and how these technologies are going to impact the veterinary world. Our next "bleeding edge" technology is something that some people say will not only revolutionise how we live our daily lives, but may also even make us immortal. It's a scary prospect to some people, but the truth is, many of us don't even know it exists yet. I'm talking, of course, about Artificial Intelligence (AI).

What is artificial intelligence?

Put simply, artificial intelligence refers to the ability of machines to exhibit human-like intelligence, acting in "intelligent" ways that demonstrate cognitive abilities such as learning, problem-solving and decision-making. Many Sci-fi films have been made on the subject of artificial intelligence, from Transcendence, where Johnny Depp's character becomes a frighteningly powerful artificially intelligent being, to the romantic drama Her, where Joaquin Phoenix's character falls in love with an AI-enhanced computer.

Needless to say, this evolution in technology is a frightening yet fascinating prospect which opens a world of new opportunities. Ethically speaking, creating machines with greater intelligence than human beings is something of a Pandora's box, but, as we will see in this chapter, it will also bring a wealth of new capabilities in the medical field which could ultimately save lives and advance human and animal medicine as we know it.

Progression

As technology advances, computers are engineered with more and more new skills. These new skills are at first heralded as examples of AI but, as they become more widely used, they lose their AI label and become normalised. An example of one of these now normal abilities is optical character recognition (OCR) or the ability of machines to read and encode the printed word.

Some examples of abilities that we currently class as artificial intelligence include:
- Understanding human speech
- Competing against humans in games like chess and Go
- Self-driving cars
- Interpreting complex data

The field of AI

One of the long term goals of the field of artificial intelligence is to create machines with a level of general intelligence that matches or surpasses that of a human. Other problems or goals that the field is concerned with include:

- Reasoning
- Planning
- Learning
- Natural language processing
- Communication
- Perception
- Moving and manipulating objects

This is a field that draws on a large number of different disciplines, including computer science, mathematics, psychology, linguistics, philosophy, neuroscience and artificial psychology.However, it's important to note that AI is not strictly a new field. In fact, ever since the inception of the computer, scientists have dreamed about creating a computerised version of the human brain. The term "artificial intelligence" was first coined at a conference at Dartmouth College in 1956.

Attendees at the conference were the leaders and founders of AI, John McCarthy, Marvin Minsky, Allen Newell, Arthur Samuel and Herbert Simon. The computer programmes they were writing at the time were nothing short of astonishing and they showed people that computers were capable of solving algebra problems, winning at checkers and even speaking English. From that point onwards, the field of artificial intelligence evolved and spread throughout the world.

Examples of AI in daily life

Another myth surrounding AI is that it exists in some imagined utopian (or dystopian) future. Most people believe it is something we are still very far from achieving, when in fact, we are already using various facets of artificial intelligence in our daily lives. AI is being used everywhere from businesses to the internet, and it is here to stay. Here are some examples: Virtual assistants like Siri, Google Home, Cortana and Amazon Echo are gaining popularity fast.

These voice activated assistants perform a number of tasks, from internet searches to making to-do lists and playing music on demand. Siri in particular has gained a lot of media coverage due to its witty responses, often used when people try to outsmart the device.

Virtual customer assistants are also becoming widespread. These pop up in chat boxes when you're using a website, asking if you need help. More often than not, these are computer generated bots rather than actual people chatting in real-time. Shell is an example of a company already using this type of AI.

Personalised news feeds such as the ones we see on Facebook and LinkedIn can be described as a form of AI. The feeds use a complex algorithm to provide personalised content tailored to your likes and interests. The algorithms are designed to find out about the user and search for more content they might be interested in. Smart cars may not be a reality just yet, but how many times have you seen them featured in the newspapers recently?

In 2015, The Washington Post reported that Google was developing an algorithm that could potentially let self-driving cars learn to drive through experience, just like human drivers do.

The AI they are using has learned to play video games, and Google is planning to test its intelligence in playing driving games before it moves to driving in real life. Tesla is another company looking into self-driving cars; its autopilot feature is already being used on the road.

iRobot Roomba may look like just another fancy vacuum cleaner, but it is actually a great example of artificial intelligence in action. The Roomba uses machine learning to detect how many rooms it can clean, and it uses automated reasoning to decipher which areas it can clean. It uses computer vision to detect dirt and even has an auto charge function to make sure it is always charged.

AI in the medical field

Medicine is an area which can benefit greatly from the abilities of artificial intelligence. With physicians often short of time and hospitals often strapped for cash and under staffed, there is a huge need in this field for devices and inventions that save time and increase expertise. There is an enormous amount of pressure on doctors (and on vets) to make judgements and decisions quickly, drawing on only the knowledge stored in their memories. Only in very rare situations are doctors called upon to check literature for validation that they have made the right diagnosis.

It is a tough occupation where the margin for error is a large one. it's only natural then, that the medical field is getting as excited as the rest of the world about the prospect of artificial intelligence and the new possibilities it brings to the table. Since computers these days are evolving to become intellectual instruments, capable of deducting and decision-making, it is only logical that they could be integrated into the structure of the medical system. Those in the medical field who are advocates of the use of AI envision that there will be frequent dialogue between doctors and computers.

Computers will continuously take in data such as medical history, laboratory results, and physical symptoms, ultimately assessing all of this to decide on a probable diagnosis and suggesting appropriate treatment. Since super computers are able to analyse huge amounts of data and cross-reference it in a fraction of the time it would take a human, these new diagnostic techniques will be an enormous boost to the medical profession.

Once it reaches the veterinary profession (and, as we will discuss later on, this is fast becoming a reality), it will make even bigger waves, allowing veterinarians across the world to share a vast database of diagnostic knowledge in a field where our furry patients can't tell us how they are feeling or what they are experiencing. Since veterinarians are also expected to know a vast amount about a wide range of different species and breeds, this kind of instantly accessible knowledge will be nothing short of a godsend.

Modernising medicine

Whilst tech giants like Amazon and Google are investing millions into creating AI products such as intuitive search engines and personal assistants, the same level of innovation is trickling down to the medical field, thanks to the availability of cheaper computers and a desire to create cheaper, easier healthcare for all.

It seems that physicians are increasingly relying on technology to aid their work, from IBM's Watson supercomputer (which we'll discuss later) to using smartphone-like pop up notifications. Powerful and large databases of medical information are being created and shared across the medical field, which are helping doctors everywhere to access a huge amount of invaluable knowledge applicable to their patient's needs.

One such repository is called Modernizing Medicine. This tool can be searched by doctors who are struggling to find the correct diagnosis or treatment for a particular case. Kaita Mariwalla, a dermatologist, told Wired magazine in 2014 that she used the web repository to diagnose a rare skin condition within seconds. "It gives you access to data, and data is king," Mariwalla told the magazine. The system allows doctors to tap into data from over 14 million patient visits, something that, until recently, would have been impossible for any medical professional to do.

Making diagnosis easier

One of the most widely talked about examples of AI in the medical field is IBM's supercomputer "Watson". In Japan, doctors were baffled as to why one female patient's leukaemia was resisting treatment. The case was mystifying doctors, so they decided to call on the powers of IBM's Watson machine to see what could be done.

This turned out to be a life-saving move. Within just ten minutes, the machine had studied the patient's medical history and had cross-referenced her condition against 20 million similar health records - all of which had been uploaded to the system by doctors at the University of Tokyo's Institute of Medical Science. What did Watson discover?

That the patient was suffering from a rare variation of leukaemia - a different form than had been diagnosed - and that this type needed a different course of treatment.

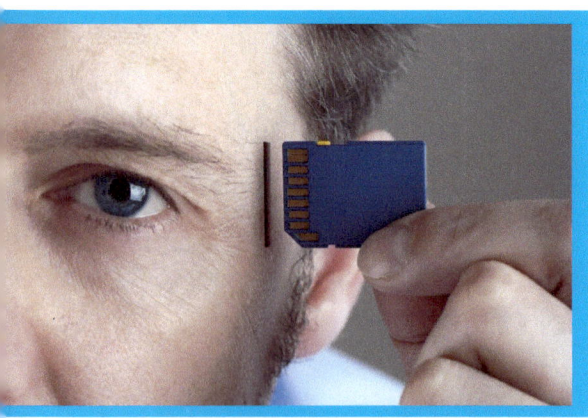

Doctors were then able to try the new treatment, with much better success this time. This just shows that supercomputers like Watson have the ability to revolutionise medicine from the inside out.

These computers can store and recall vast amounts of medical information, from every medical journal article on a particular illness to symptoms, real-life case studies and treatments. With this kind of information at our fingertips, mis-diagnosis will virtually become a thing of the past, and many lives will be saved through these new diagnostic tools.

Watson was originally programmed as a Q&A tool or a question answering computer, whch was capable of answering questions posed in natural language. Scientists configured the computer to answer questions on the game show Jeopardy!, and in 2011 the machine competed against two human contestants and won the $1 million prize. How did Watson win?

At the time, it had access to 200 million pages of structured and unstructured pages of content, including the entire text of Wikipedia, although at the time of the game it was not connected to the internet. Other sources that Watson draws upon include encyclopaedias, dictionaries, thesauri, newswire articles, and literary works. It also used databases, taxonomies, and ontologies.

In 2013, IBM announced that Watson would have its first commercial application, in the medical field. It would be used to make decisions in lung cancer treatment at Memorial Sloan Kettering Cancer Center, New York City. Since then, it has gone on to use its machine learning abilities to make numerous decisions and diagnostic observations like the Japanese cancer case described above.

IBM's website says that Watson's abilities go far beyond this, though, allowing us to compare data in nearly every aspect of our daily lives. The company's Watson video claims that:
* 415 million diabetes patients will soon be able to predict attacks before they occur
* 9 million patients can benefit from personalized cancer care
* 2.2 billion weather sensors are helping predict disasters
* 10 million tons of CO_2 emissions are being reduced

The veterinary sector

If you're looking forward to the day when Watson is finally used in the veterinary world, you can stop waiting. A version of Watson, known as Sofie, is already being used for veterinary purposes. Sofie is just one of many "apps" that are being developed to use the Watson model in various research and business fields, including travel, retail, banking and, of course, medicine. These projects are spread across six continents around the world, including Canada, where a company called LifeLearn Inc has been developing the app for the veterinary market.

Mark Stephenson, a veterinarian who also heads up development for LifeLearn, told Toronto news site The Star in 2014: "The app was developed to address a number of really significant challenges in the veterinary space. We have to deal with all species and all body systems, and we're often taking on multiple roles in the clinic besides medicine even such as finance, HR and marketing."

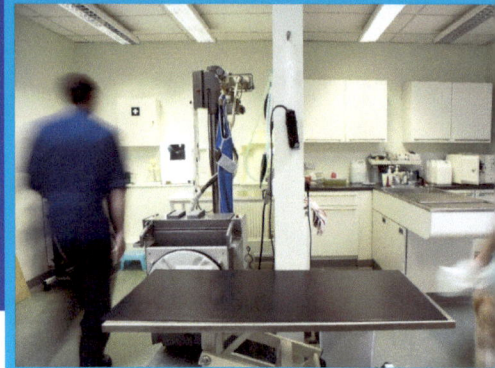

The app, or "Sofie" as it is also known, will save busy, multi-tasking veterinarians a huge amount of time-consuming research, and will also improve the accuracy of their work. It will do this by searching a vast number of research journals and textbooks, along with other relevant literature, and will return responses within seconds. Vets will not only be able to see one response, but will be given a series of responses so that they can see the true breadth and depth of the research available.

As Stephenson says, often a vet will suspect a disease such as pancreatitis, and they will want to know the fastest and most reliable test that will confirm their suspicions, for the particular species they are dealing with. Sofie will provide quick, reliable answers for these kinds of questions, at the exact moment they are needed. It will also be invaluable for emergency and critical care situations, where time is of the essence.

The team at LifeLearn have been teaching Sofie about all branches of veterinary medicine, but she will also, due to her machine learning abilities, learn and develop her knowledge each time she is used. The new technology won't replace veterinarians, as such, but will act like a colleague they can turn to for advice whenever they need to. Sofie's abilities are far beyond those of a normal computer or a simple search engine.

Firstly, her language processing ability allows her to communicate with veterinarians on a human level, without the need for a go-between to decode her responses. She doesn't just serve up content, either; she can actually read the content and figure out how that content relates to other pieces of content.

This allows her to find and make connections between, for example, rare symptoms and likely diagnoses or treatments. As well as all of this, Sofie can handle and process enormous volumes of data, in a field where that data is growing all the time and researchers are constantly making new discoveries about animals and their treatment.

Lastly, the wonderful thing about Sofie and her "parent" Watson is that these supercomputers are now considerably smaller and more compact than when they first started out, due to the fact that their "brains" can be stored in the cloud. The cloud allows for an almost unlimited amount of data to be stored.

LifeLearn has reported that, as of May 2015, over 200 users across 57 veterinary hospitals are using Sofie. As such, we can safely say that it won't be too long before it becomes the norm, and not the exception, to use these supercomputers to help our four legged friends. When this happens, our pets will have a higher survival rate and better access to expert opinions that could save their lives.

Life for our vets will be easier too; instead of relying on trial and error, they'll be able to use a supercomputer to look up the most effective treatments for a particular disease, and will be able to cross reference odd or rare symptoms with a large database of medical case studies. With the click of a button, every vet will have access to specialist information on every species known to man, thus revolutionising the field of veterinary medicine.

CONCLUSION

So, what are we to conclude about the future of veterinary medicine? One thing we know for certain is that the field itself is evolving quickly. The job of a veterinarian is growing and changing as we speak, and is expanding to include new specialisms and techniques. Many of these new "bleeding edge" technologies will make veterinarian's lives easier, which is excellent news for a profession where a single person often has to be a cardiologist, dermatologist, dentist, and radiologist all at once. In this book, we have seen various areas in which we can expect huge changes over the coming decades. Some are in their infancy, and some are already underway, set to change how we do veterinary medicine very soon. But how will these techniques change veterinary medicine in general?

Let's take a look at the bigger picture: Vets will start to learn organ transplant techniques as well as how to design prosthetics.

In chapter one, we talked about 3D printing and how this will change animal's lives in two key areas: prosthetics and printed organs. Prosthetics are already being printed, but entire replicas of organs are going to take a lot longer to master. Nonetheless, with a new supply of organs that don't rely on living donors, we can expect the practice of transplants to become the norm for those who can afford the very best treatments for their pets. Since organ transplants don't happen very often in the veterinary world, we can see these new changes as an expansion of the veterinary field and an addition to the already huge range of specialty treatments that a veterinarian might be expected to perform.

Vets will rely on robots for specific surgeries

In chapter two, we discussed the field of robotics and how a new breed of robot-surgeons will change how veterinary surgeons operate forever. With infection rates significantly reduced and the margin for human error also reduced, we can expect fantastic things from these new robotic techniques. These devices will see the role of the veterinarian evolve into an overseeing capacity, where the maintenance and operation of these robots will be just as important as the surgeries themselves. Veterinary surgeon's daily jobs may be less "creative" in terms of solving the various problems of surgery, but they will also be less plagued by fatigue and the stress of human error.

Vets will use virtual reality to help train for surgery, and augmented reality as an everyday tool

In chapter three, we discussed virtual reality and augmented reality. These two different disciplines will make training to be a vet more interactive and perhaps easier than ever before. Trainees will see 60 degree views of virtual surgeries, during which they will be able to pause the action, rewind, or inspect various tools and elements. With augmented reality, practising vets may be able to get instant data in their field of vision, simply be looking at their four legged patients. These developments will ultimately make the veterinary field a more dynamic, progressive profession where technology makes learning and data more accessible to all. Vets will rely on supercomputers to help them make important medical decisions

In chapter four, we talked about artificial intelligence as a new way for veterinarians to make important diagnostic and treatment decisions. These new supercomputers will hold vast amounts of veterinary information. Vets will no longer need to refer to huge textbooks of information when they need advice, and they won't need to call in a specialist when they need a second opinion. All of the information they need will be available in seconds, at the touch of a button. They will also be able to share and find information on rare conditions and rare species, sharing knowledge with vets across thousands of miles and improving the overall quality and efficiency of the veterinary field.

With all of these huge changes happening in the veterinary world, it's important to note that we aren't going to be taking our pets to robotic vets any time soon. Even though they'll be relying on supercomputers and robotic surgeries, vets will still be the ones who make the final decisions and call the shots – we will always need this human element in medicine. The veterinary profession isn't going anywhere any time soon – in fact, as we have seen in this book, it is about to become more interesting, and more diverse, than ever before.

REFERENCE

Chapter 1

- BSR issue brief: 3-D Printing Sustainability Opportunities and Challenges, November 2015 https://www.bsr.org/our-insights/report-view/3d-printing-report
- Deloitte: Disruptive Manufacturing: the effects of 3D printing https://www2.deloitte.com/content/dam/Deloitte/ca/Documents/insights-and-issues/ca-en-insights-issues-disruptive-manufacturing.pdf
- 3D Printing: The Next Revolution in Industrial Manufacturing - new research from UPS and the Consumer Technology Association https://www.ups.com/media/en/3D_Printing_executive_summary.pdf
- Enabling the Future http://enablingthefuture.org/about/
- Lucky dog runs free with 3D printed prosthetics https://www.cnet.com/news/lucky-dog-back-on-its-feet-with-3d-printed-front-leg-prosthetics/
- The Future of 3D Printed Prosthetics https://techcrunch.com/2016/06/26/the-future-of-3d-printed-prosthetics/
- Beauty and the Beak http://www.birdsofpreynorthwest.org/beauty-and-the-beak.html
- The Next Frontier in 3D Printing: Human Organs http://edition.cnn.com/2014/04/03/tech/innovation/3-d-printing-human-organs/
- The History of 3D Printing from its inception to today https://www.preceden.com/timelines/71092-3d-printing
- Organ Printing https://en.wikipedia.org/wiki/Organ_printing

Chapter 2

- Robotic Surgery: The Ultimate Surgeon, Zander Brais http://www.odec.ca/projects/2008/brai8z2/Report.pdf
- Robotics in Medical Applications, Jackrit Suthakorn, Ph.D. http://www.bartlab.org/Dr.%20Jackrit's%20Papers/international%20proceedings/Robotics%20in%20Medical%20Application.pdf

- The Medical Post, volume 21, no 23 http://www.brianday.ca/imagez/1051_28738.pdf
- Background and History of Surgical Robotics http://al-laboutroboticsurgery.com/roboticsurgeryhistory.html
- TED talk: Surgery's past, present and robotic future https://www.ted.com/talks/catherine_mohr_surgery_s_past_present_and_robotic_future
- Da Vinci Surgical System https://en.wikipedia.org/wiki/Da_Vinci_Surgical_System
- Da Vinci Surgical System http://www.davincisurgery.com/da-vinci-surgery/da-vinci-surgical-system/
- Robot Performs Surgery on a Lion https://totallycoolpix.com/magazine/2015/05/robot-performs-surgery-on-a-lion
- Amazing Ways Robots are Being Used in Medicine http://infinigeek.com/5-amazing-ways-that-robots-are-being-used-in-medicine/
- 10 Medical Robots that Could Change Healthcare http://www.informationweek.com/mobile/10-medical-robots-that-could-change-healthcare/d/d-id/1107696?page_number=1
- Using Nanotechnology in Medicine http://www.yalescientific.org/2013/02/microbots-using-nanotechnology-in-medicine/
- Nanotechnology and the life-saving future of medicine https://www.theguardian.com/what-is-nano/nano-and-the-life-saving-future-of-medicine

Chapter 3

- 9 Healthcare Companies Making Innovations in Virtual Reality https://touchstoneresearch.com/the-9-healthcare-companies-making-innovations-in-virtual-reality/
- Cognizant report: Disrupting Reality: Taking Virtual & Augmented Reality to the Enterprise https://www.cognizant.com/whitepapers/disrupting-reality-taking-virtual-augmented-reality-to-the-enterprise-codex2124.pdf
- Augmented Reality, the Future and Pokemon Go http://www.forbes.com/sites/kevinanderton/2016/11/14/augmented-reality-the-future-and-pokemon-go-infographic/#7e-37033a4e66

- How Google Glass can Help Veterinarians http://www.veterinarypracticenews.com/Vet-Practice-Videos/Google-Glass-Uw-School-Veterinary-Medicine/
- Augmented Reality Brings a New Dimension to Client Education https://www.bva.co.uk/professional-development/vets-tv/veterinary-view/augmented-reality-brings-a-new-dimension-to-client-education---virbac/
- Smartphones Transform Veterinary Training http://www.horseandhound.co.uk/news/smartphones-transform-veterinary-training-470200#RHMX1d1ZMLOiUOPW.99
- 6 Cool Uses for Augmented Reality in Healthcare http://www.techrepublic.com/article/6-cool-uses-for-augmented-reality-in-healthcare/

Chapter 4

- Artificial Intelligence: Trends and Predictions for 2030 https://www.qulix.com/wp-content/uploads/2016/10/Artificial_intelligence.pdf
- Incredible Examples of Artificial Intelligence Online So Far https://blog.digital22.com/7-incredible-examples-of-artificial-intelligence-online-so-far
- Examples of Artificial Intelligence http://beebom.com/examples-of-artificial-intelligence/
- Artificial Intelligence https://en.wikipedia.org/wiki/Artificial_intelligence
- Artificial Intelligence in Medicine https://healthinformatics.wikispaces.com/Artificial+Intelligence+in+Medicine
- Artificial Intelligence is Now Telling Doctors How to Treat You https://www.wired.com/2014/06/ai-healthcare/
- LifeLearn's Dr Sofie Aims to Know All The Answers http://www.veterinarypracticenews.com/LifeLearns-Dr-Sofie-Aims-to-Know-All-the-Answers/
- IBMs Watson Computer Goes to Veterinary School https://www.thestar.com/news/insight/2014/10/12/ibms_watson_computer_goes_to_veterinary_school.html

ABOUT THE AUTHOR

Dr Gordon Roberts has had a lifelong love affair with the future of healthcare.

Growing up as a child in New Zealand, he would make regular trips to a nearby forest and, surrounded by some of the Southern Hemisphere's most striking landscapes, he would spend his time wondering what future vet care would look like.

Even then, he had a strong feeling that what we consider to be an accepted standard in healthcare would be replaced by an entirely new paradigm.

He saw a world where the lives of both humans and animals would be greatly improved and extended by the advances to come. This forward-thinking start in life led him to where he is today, one of the world's leading futurist veterinarians with a passion for the medicine of tomorrow.

He is particularly interested in the intersection between conscious energy, medicine and technology. Inspired by exponential advances in human medicine, Gordon is helping bring this knowledge to veterinary science and the broader pet welfare industry.

Gordon divides his time between his native New Zealand (where he lives with his wife, four children and many pets) and confer- ences, seminars and airport terminals around the world as he spreads the word about futurist vet opportunities, while dealing with his angel investments in future tech.

Want to read more about these exciting developments as they happen? All the latest news and discoveries from the future of veterinary medicine can be found on Gordon's website futuristvet.com

www.ingramcontent.com/pod-product-compliance
Lightning Source LLC
Chambersburg PA
CBHW040844180526
45159CB00001B/306